Health, safety and risk

Looking after each other at school and in the world of work

Written by Dorothy Warren
RSC School Teacher Fellow 1999–2000

ROYAL SOCIETY OF CHEMISTRY

Health, safety and risk

Written by Dorothy Warren

Edited by Colin Osborne and Maria Pack

Designed by Imogen Bertin

Published and distributed by Royal Society of Chemistry

Printed by Royal Society of Chemistry

Copyright © Royal Society of Chemistry 2001

Registered charity No. 207980

For further information on other educational activities undertaken by the Royal Society of Chemistry write to:

Education Department
Royal Society of Chemistry
Burlington house
Piccadilly
London W1J 0BA

Information on other Royal Society of Chemistry activities can be found on its websites:
http://www.rsc.org
http://www.chemsoc.org
http://www.chemsoc.org/LearnNet contains resources for teachers and students from around the world.

ISBN 0–85404–959–2

British Library Cataloguing in Publication Data.

A catalogue for this book is available from the British Library.

RS•C

Foreword

Young people today grow up in a world where the media bombard them with information about the effect of various factors on their health, the safety of everything from household items to nuclear waste, and the risks associated with them. These ideas of health, safety and risk are often poorly understood. The Society has produced this resource in order to help teachers bring about a better understanding of these concepts in their students and so that, as young people growing up in a scientific and technological society, they can be properly aware of the risks associated with the world they live in, and of attempts to minimize them.

Professor Steven Ley CChem FRSC FRS
President, The Royal Society of Chemistry

RS•C

Acknowledgements

The production of this book was only made possible because of the advice and assistance of a large number of people. To the following, and everyone who has been involved with this project, including the members of the science staff and students in trial schools, both the author and the Royal Society of Chemistry express their gratitude.

General

Colin Osbome, Education Manager, Schools & Colleges, Royal Society of Chemistry
Maria Pack, Assistant Education Manager, Schools & Colleges, Royal Society of Chemistry
Members of the Royal Society of Chemistry Committee for Schools and Colleges.
Members of University of York Science Education Group.
Jill Bancroft, Special educational needs project officer, CIEC
Donald Stewart, Dundee College, Dundee
Richard Warren, Mathematics Department, Ampleforth Colledge, York
Bob Campbell, Department of Educational Studies, University of York
Professor David Waddington, Department of Chemistry, University of York.
Eric Albone of the Clifton Scientific Trust, Bristol
Peter Borrows, CLEAPSS School Science Service, Brunel University
Bronwen Freake, Greenbank Residential School, Hartford.
National Radiation Protection Board
Bristol and District Local Section of the RSC
Peter Dawson, Science adviser, York and the York Schools Secondary science group.

Schools

David Billett, Ampleforth College, York
Howard Campion, Fulford School, York.
Gavin Cowley, Oaklands School, York
Tim Gayler, Little Ilford School, London
Carole Lowrie, Hummersknott School & Language College, Darlington
Simon Howard, Ampleforth College, York.

The Royal Society of Chemistry would like to extend its gratitude to the Department of Educational Studies at the University of York for providing office and laboratory accommodation for this Fellowship and the Head Teacher and Governors of Fulford Comprehensive School, York for seconding Dorothy Warren to the Society's Education Department.

RS•C

Contents

How to use this resource iv

Introduction 1

A risky life 2
   Teachers' notes 2
   Answers 5
   Student worksheets – photocopiable masters
   Our perception of risk 7
   The risk and dread exercise 8
   Who is safest at school? 11
   Risk assessments 12

Safety symbols 14
   Teachers' notes 15
   Answers 16
   Student worksheets – photocopiable masters
   Safety symbols 1 17
   Safety symbols 2 19

Assessing safety in science experiments 20
   Teachers' notes 20
   Answers 21
   Student worksheets – photocopiable masters
   Assessing risk in science experiments 22

Risk benefit analysis 25
   Teachers' notes 25
   Student worksheets – photocopiable masters
   Risk benefit analysis 27

Alice Hamilton – safety, hazards and risk 28
   Teachers' notes 28
   Answers 31
   Student worksheets – photocopiable masters
   Alice Hamilton investigates public health 33
   Carbon monoxide: introducing the problems 37
   Carbon monoxide project work 39
   A detector calls – article from *Chemistry in Britain* 40

The nitrogen oxide story 43
   Teachers' notes 43
   Answers 43
   Student worksheets – photocopiable masters
   The nitrogen oxide story 44

Radiation doses 46
   Teachers' notes 46
   Answers 48
   Student worksheets – photocopiable masters
   Radiation doses 51

References 56

RS•C

How to use this resource

At the start of the 21st century secondary education yet again underwent changes. These included the introduction of new curricula at all levels in England, Wales and Scotland and the Northern Ireland National Curriculum undergoing review. With more emphasis on cross curricula topics such as health, safety and risk, citizenship, education for sustainable development, key skills, literacy, numeracy and ICT, chemistry teachers must not only become more flexible and adaptable in their teaching approaches, but keep up to date with current scientific thinking. The major change to the science 11–16 curricula of England and Wales was the introduction of 'ideas and evidence in science', as part of Scientific Enquiry. This is similar to the 'developing informed attitudes' in the Scottish 5–14 Environmental studies, and is summarised in Figure 1.

In this series of resources, I have attempted to address the above challenges facing teachers, by providing:

■ A wide range of teaching and learning activities, linking many of the cross-curricular themes to chemistry. Using a range of learning styles is an important teaching strategy because it ensures that no students are disadvantaged by always using approaches that do not suit them.

■ Up-to-date background information for teachers on subjects such as global warming and Green Chemistry. In the world of climate change, air pollution and sustainable development resource material soon becomes dated as new data and scientific ideas emerge. To overcome this problem, the resources have been linked to relevant websites, making them only a click away from obtaining, for example, the latest UK ozone data or design of fuel cell.

■ Resources to enable ideas and evidence in science to be taught within normal chemistry or science lessons. There is a need to combine experimental work with alternative strategies, if some of the concerns shown in Figure 1, such as social or political factors, are to be taught. This can be done for example, by looking at the way in which scientists past and present have carried out their work and how external factors such a political climate, war and public opinion, have impinged on it.

■ Activities that will enhance student's investigative skills.

These activities are intended to make students think about how they carry out investigations and to encourage them to realise that science is not a black and white subject. The true nature of science is very creative, full of uncertainties and data interpretation can and does lead to controversy and sometimes public outcry. Some of the experiments and activities will be very familiar, but the context in which they are embedded provide opportunities for meeting other requirements of the curriculum. Other activities are original and will have to be tried out and carefully thought through before being used in the classroom. Student activities have been trialled in a wide range of schools and where appropriate, subsequently modified in response to the feedback received.

Dorothy Warren

RS•C

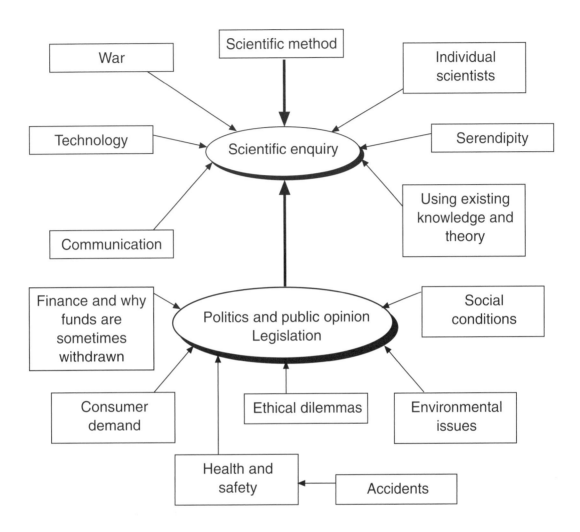

Figure 1 The factors influencing the nature of scientific ideas – scientific enquiry and the advancement of science

Maximising the potential use of this resource

It is hoped that this resource will be widely used in schools throughout the United Kingdom. However, as every teacher knows, difficulties can be experienced when using published material. No single worksheet can cater for the needs of every student in every class, let alone every student in every school. Therefore many teachers like to produce their own worksheets, tailored to meet the needs of their own students. It was not very surprising when feedback from trial schools requested differentiated worksheets to allow access to students of different abilities. In an attempt to address these issues and concerns, this publication allows the worksheet text and some diagrams to be modified. All the student worksheets can be downloaded in Word format, from the Internet via the LearnNet website, **http://www.chemsoc.org/networks/learnnet/ideas-evidence.htm** .This means that the teacher can take the basic concepts of the activity, and then adapt the worksheet to meet the needs of their own students. Towards the end of the teachers' notes for most activities there are some suggestions as to how the resource can be adapted to meet the needs of students of different abilities. There are also some examples of differentiated worksheets included in the resource.

RS•C

It is not envisaged that teachers will use every activity from each piece of work with an individual class, but rather pick and choose what is appropriate.

Activities that involve researching for secondary information on the Internet contain hyperlinks to appropriate websites. To minimise the mechanical typing of the URLs and possible subsequent errors, the students can be given the worksheet in electronic form and asked to type in their answers. The websites are then only a click away.

Appropriate secondary information has been included in the teachers' notes for use in class when the Internet or ICT room is unavailable.

Unfortunately, from time to time website addresses do change. At the time of publication all the addresses were correct and the date that the site was last accessed is given in brackets. To minimise the frustration experienced when this happens, it is advisable to check the links before the lesson. If you find that a site has moved, please email both **LearnNet@rsc.org** and **education@rsc.org** giving full details so that the link can be updated on the worksheets on the web in the future.

Strategies for differentiated teaching

All students require differentiated teaching and it is not just an issue for those students with special educational needs. The following definition by Lewis[1] has been found to be quite useful.

'Differentiation is the process of adjusting teaching to meet the needs of individual students.'

Differentiation is a complex issue and is very hard to get right. It can be involved in every stage of the lesson *ie* during planning (differentiation by task), at the end of the activity (differentiation by outcome) and ongoing during the activity. Often teachers modify the activity during the lesson in response to feedback from the class. Differentiation does not only rely on appropriate curriculum material but is also concerned with maximizing learning. Student involvement and motivation effect the learning experience and should be considered and taken into account. It is therefore not surprising that differentiation is one of the areas of classroom teaching where teachers often feel under-confident. Most strategies for differentiated lessons are just applying good teaching practice *eg* varying the pace of the lesson, providing suitable resources and varying the amount and nature of teacher intervention and time.[2] Rather than just providing several examples of differentiated activities from the same worksheet, a list of strategies for differentiated teaching is presented, with some examples of how they can be used in the classroom. The examples can be found at the appropriate places in the text.

1. Using a range of teaching styles
A class is made up of different personalities, who probably have preferred learning styles. Using a range of teaching approaches makes it more likely that all students will be able to respond to the science that is being taught. The following examples have been included and can be found at the appropriate place in the resource.

For example, the student worksheet **Risk assessments** can be used either as a stand-alone resource or as a basis for group discussion or role-play.

2. Varying the method of presentation or recording
Giving the students some choice about how they do their work. There are many opportunities given throughout the resource.

3. Taking the pupil's ideas into account

Provide opportunities for students to contribute their own ideas to the lesson. For example when setting up an investigation allow different students the freedom to choose which variables they are going to investigate. The use of concept cartoons provides an ideal opportunity for students to discuss different scientific concepts. (See D. Warren, *The nature of science*, London: Royal Society of Chemistry, 2001.)

4. Preparing suitable questions in advance

Class discussions are important in motivation, exploring ideas, assessment etc. Having a list of questions of different levels prepared in advance can help to push the class.

5. Adjusting the level of scientific skills required

Example – using symbol equations or word equations

6. Adjusting the level of linguistic skills required

Example – safety symbols

Safety symbols 1 – a high reading age

Safety symbols 2 – a low reading age

Teachers may like to check the readability of their materials and of the texts they use. Guidance on this and on the readability of a range of current texts may be found at **http://www.timetabler.com/contents.html** (accessed June 2001).

References

1. A. Lewis, *British Journal of Special Education*, 1992, **19**, 24–7.

2. S. Naylor, B. Keogh, *School Science Review*, 1995, **77(279)**, 106–110.

How scientists communicate their ideas

Effective communication is crucial to the advancement of science and technology. All around the globe there are groups of research scientists and engineers, in universities and in industry, working on similar scientific and technological projects. Communication between these groups not only gives the scientist new ideas for further investigations, but helps in the evaluation of data. Results from different groups will either help to confirm or reject a set of experimental data. Communication is vital when a company wants to sell a new product. Depending on the product the buyer will want to understand how it works and how to maintain it. Several of the employees will have to learn how to use the product, and respond quickly to changing technology and circumstances. Therefore the manufacturers must be able to communicate the science to prospective buyers.

Scientists communicate in a number of ways including:

- Publication in research journals

- Presenting papers at scientific conferences

- Poster presentations at conferences

- Book reviews by other scientists

- Publication on the Internet

- Sales brochures

- Advertising flyers

- Television documentaries.

RS•C

Publication in research journals

The article is written. The article must have an abstract, which is a short summary.

It is submitted to a journal.

The article is refereed by other scientists, working in a similar area. This is to check that the work is correct and original.

The article may be returned to the author to make changes.

The article is accepted and published by the journal.

The article is published.

Presenting papers at scientific conferences

Conference organisers invite scientists to speak on specific topics and projects.

An abstract is submitted to and accepted by the conference organisers.

The conference programme is organised and the speakers notified.

The scientist gives their talks, usually aided by slides or multimedia presentations, which contain the main points.

There is usually time for questions after the talk.

The written paper is given to the conference organisers.

All the papers are published in the conference proceedings. This is usually a book.

Poster presentations at conferences

An abstract is submitted to and accepted by the conference organisers.

The conference programme is organised and the poster people notified.

During the poster session the authors stand by the posters, ready to answer any questions as the delegates read the posters.

Written papers may then be published in the conference proceedings.

Book reviews

Other scientists in the same field often review new books. The reviews are then published in scientific magazines and journals. The review offers a critical summary of the book. The idea of the review is to give possible readers an idea of the contents and whether it is suitable for the intended purpose.

Publishing on the Internet

This is the easiest way to publish. Anyone can create their own web page and publish their own work. In this case the work is not refereed or checked by other people.

However, a lot of the information published on the Internet is linked to reputable organisations. In this case the articles will have been checked before they are published. Much of the information published on the Internet is targeted at the general public, and therefore the scientific ideas are presented in a comprehensible way. There are often chat pages so people can communicate their views and ask questions or request further information. The power of the Internet is that there is the opportunity to get immediate feedback to a comment or question.

Sales brochures

The information must be presented in an attractive and concise manner. After all you are trying to sell something. There should be a balance between technical information and operating instructions!

RS•C

Advertising flyers

This must be written with the target audience in mind.

The information must be concise as there is limited space. The format must be attractive and should include pictures as well as writing. The flyer should also be quite cheap to produce.

Teaching students to communicate ideas in science

Students can be taught effective communication skills:

■ By encouraging communication between students and a range of audiences in classrooms

■ By encouraging them to investigate like 'real scientists' by reporting their findings for checking and testing by others, and participating in two-way communication. (Communicating between groups, classes, partner schools, schools abroad perhaps via the Internet.)

■ By setting investigations in a social context which offers the opportunity to communicate the project outside of the classroom. These work best when there is local interest.

When presenting investigative work to an audience, the student should consider the following:

■ Who will be in the audience?

■ What information does the audience need to know *eg* method, results and recommendations?

■ How to present the information in an interesting and professional way *eg* should graphs be hand drawn or done on the computer?

■ That the information offered convinces the audience that their investigation was valid and reliable.

■ Poster presentations or display boards should be concise, since the space is limited.

■ When speaking to audiences remain calm, speak clearly and slowly and try to be enthusiastic. Make sure that information on slides and OHTs can be read from the back of the room.

When writing a report of the findings of a scientific investigation for others to check and test, the emphasis should be on clarity. Another person is going to carry out the same investigation. The only information available is what is written in the report.

The report could be written under the following headings:

■ Introduction

■ Scientific knowledge

■ Planning

■ Table of results

■ Graphs

■ Conclusions

■ Evaluation

■ Recommendations.

RS•C

Further background information

R. Feasy, J. Siraj-Blatchford, *Key Skills: Communication in Science,* Durham: The University of Durham / Tyneside TEC Limited, 1998.

Curriculum coverage

Curriculum links to activities in this resource are detailed at
http://www.chemsoc.org/networks/learnnet/hsr.htm

Curriculum links to activities in other resources in this series are detailed at

http://www.chemsoc.org/networks/learnnet/ideas-evidence.htm

Health and safety

All the activities in this book can be carried out safely in schools. The hazards have been identified and any risks from them reduced to insignificant levels by the adoption of suitable control measures. However, we also think it is worth explaining the strategies we have adopted to reduce the risks in this way.

Regulations made under the Health and Safety at Work etc Act 1974 require a risk assessment to be carried out before hazardous chemicals are used or made, or a hazardous procedure is carried out. Risk assessment is your employers responsibility. The task of assessing risk in particular situations may well be delegated by the employer to the head of science/chemistry, who will be expected to operate within the employer's guidelines. Following guidance from the Health and Safety Executive most education employers have adopted various nationally available texts as the basis for their model risk assessments. These commonly include the following:

Safeguards in the School Laboratory, 11th edition, ASE, 2001

Topics in Safety, 3rd Edition, ASE, 2001

Hazcards, CLEAPSS, 1998 (or 1995)

Laboratory Handbook, CLEAPSS, 1997

Safety in Science Education, DfEE, HMSO, 1996

Hazardous Chemicals. A manual for science educaton, SSERC, 1997 (paper).

Hazardous Chemicals. An interactive manual for science education, SSERC, 1998 (CD-ROM)

If your employer has adopted more than one of these publications, you should follow the guidance given there, subject only to a need to check and consider whether minor modification is needed to deal with the special situation in your class/school. We believe that all the activities in this book are compatible with the model risk assessments listed above. However, teacher must still verify that what is proposed does conform with any code of practice produced by their employer. You also need to consider your local circumstances. Is your fume cupboard reliable? Are your students reliable?

Risk assessment involves answering two questions:

■ How likely is it that something will go wrong?

■ How serious would it be if it did go wrong?

How likely it is that something will go wrong depends on who is doing it and what sort of training and experience they have had. In most of the publications listed above there

RS•C

are suggestions as to whether an activity should be a teacher demonstration only, or could be done by students of various ages. Your employer will probably expect you to follow this guidance.

Teachers tend to think of eye protection as the main control measure to prevent injury. In fact, personal protective equipment, such as goggles or safety spectacles, is meant to protect from the unexpected. If you expect a problem, more stringent controls are needed. A range of control measures may be adopted, the following being the most common. Use:

■ a less hazardous (substitute) chemical;

■ as small a quantity as possible;

■ as low a concentration as possible;

■ a fume cupboard; and

■ safety screens (more than one is usually needed, to protect both teacher and students).

The importance of lower concentrations is not always appreciated, but the following table, showing the hazard classification of a range of common solutions, should make the point.

Ammonia (aqueous)	irritant if ≥ 3 M	corrosive if ≥ 6 M
Sodium hydroxide	irritant if ≥ 0.05 M	corrosive if ≥ 0.5 M
Ethanoic (acetic) acid	irritant if ≥ 1.5 M	corrosive if ≥ 4 M

Throughout this resource, we make frequent reference to the need to wear eye protection. Undoubtedly, chemical splash goggles, to the European Standard EN 166 3 give the best protection but students are often reluctant to wear goggles. Safety spectacles give less protection, but may be adequate if nothing which is classed as corrosive or toxic is in use. Reference to the above table will show, therefore, that if sodium hydroxide is in use, it should be more dilute than 0.5 M (M = mol dm^{-3}).

CLEAPSS Student Safety Sheets

In several of the student activities CLEAPSS student safety sheets are referred to and recommended for use in the activities. In other activities extracts from the CLEAPSS sheets have been reproduced with kind permission of Dr Peter Borrows, Director of the CLEAPSS School Science Service at Brunel University.

Teachers should note the following points about the CLEAPSS student safety sheets:

■ Extracts from more detailed student safety sheets have been reproduced.

■ Only a few examples from a much longer series of sheets have been reproduced.

■ The full series is only available to member or associate members of the CLEAPSS School Science Service.

■ At the time of writing, every LEA in England, Wales and Northern Ireland (except Middlesbrough) is a member, hence all their schools are members, as are the vast majority of independent schools, incorporated colleges and teacher training establishments and overseas establishments.

■ Members should already have copies of the sheets in their schools.

■ Members who cannot find their sheets and non-members interested in joining should contact the CLEAPSS School Science Service at Brunel University, Uxbridge,

RS•C

UB8 3PH; tel. 01895 251496; fax. 01895 814372; email science@cleapss.org.uk or visit the website **http://www.cleapss.org.uk** (accessed June 2001).

■ In Scotland all education authorities, many independent schools, colleges and universities are members of the Scottish Schools Equipment Resource Centre (SSERC). Contact SSERC at St Mary's Building, 23 Holyrood Road, Edinburgh, EH8 8AE; tel: 0131 558 8180, fax: 0131 558 8191, email: sts@sserc.org.uk or visit the website at **http://www.sserc.org.uk** (accessed June 2001).

RS•C

RS•C

Introduction

This book contains a series of student activities intended to promote health and safety within the world of school and work. The student activities place safety within the wider context of the risks we take every day. By looking back at the work of Alice Hamilton at the start of the 20th century, it is hoped that the concept of risk assessment and control will become more meaningful to students and that they will see health and safety as necessary for their own protection as well as that of others. Students should appreciate that they must take responsibility for their own health and safety and that in some circumstances, *eg* rock climbing, the risk of injury greatly increases if the correct equipment is not used.

Health, safety and risk are often to be found at the heart of public outrage and controversy. Public perception is often poor and so when topics such as Bovine Spongiform Encephalopathy (BSE), new variant Creuzfeld Jakob Disease (nvCJD) or Genetically Modified (GM) foods hit the headlines, the response from politicians, the media and the commercial world can be extreme. People are looking for immediate answers to the problems and are too impatient to wait for the results of scientific investigations.

Fitting health and safety into the science curriculum

Health, safety and risk education should be treated as an integral part of the science curriculum. Students should be taught that health and safety is an important part of the scientific method. Whenever scientific investigations are carried out at school or in industry, risk assessment is an important part of planning. If the risk is too great, should the experiment still be carried out?

Science can be used as a context to teach about the risks and hazards involved in:

- Personal and social life (*eg* as discussed in PSE lessons)

- Health and safety at work

- Environmental safety

- Food safety

- Technology and our changing world.

RS•C

A risky life

Teachers' notes

Objectives
- For students to investigate their perception of risk.
- To understand the difference between hazard and risk.
- To carry out a risk assessment of the school laboratory.
- To recognize hazards, assess consequent risks, and take steps to control the risks to themselves and others.

Outline
The teaching material is divided into three short activities that allow students to investigate their perception of risk. The activities vary in difficulty and are independent of each other. The final activity involves carrying out a risk assessment of the school laboratory and devising some rules to ensure that science lessons are taught in a safe environment.

Teaching topics

These activities are suitable for 11–14 year olds. However, many teachers may feel that it is appropriate to include some of them as part of an introduction to science for 11–12 year olds. The perception exercises can be used at various stages to provide contexts to illustrate the general concepts of health, safety and risk.

The risk assessment activity could be used as an introduction to safety symbols.

Before reading the notes below look at the student worksheets.

Background information

Today there are regulations and details concerning health and safety at work. There are many regulations concerning the 'Control of Substances Hazardous to Health' (COSHH). One of the requirements of COSHH is to reduce the health risks in the workplace, so that the risk of exposure to hazardous or poisonous substances is minimized.

Definitions

- A hazard is a source of potential loss or danger.
- A hazard is anything that can cause harm.
- Examples of hazards include a wet floor, a ladder against a wall or some chemicals.
- A risk is the possibility of suffering loss or harm.
- A risk is the chance that someone will be harmed by the hazard.

Sources of information

Health and Safety Executive **http://www.hse.gov.uk** (accessed June 2001)

Five steps to risk assessment, HSE Books, 1998 (ISBN 0717615650)

COSHH a brief guide to the regulations, HSE Books, 1999 (ISBN 0717624447)

RS•C

The two above HSE booklets are available on-line and can accessed via the HSE books website which lists all of the HSE publications, and gives details of how to order copies.

http://hsebook.co.uk/ (accessed June 2001).

Further accident statistics can be obtained from the Royal Society for the Prevention of Accidents, via their website **http://rospa.co.uk/** (accessed June 2001). The website also contains useful safety guidelines, such as information for teachers going on school trips. Case studies are available from April 2002 at **http://www.risk-ed.org**.

Teaching tips

Our perceptions of risk, **The risk and dread exercise** and **Who is safest at school?** are intended to get the class thinking about the hazards and risk that are met in everyday life and promote discussion, before learning about the more specific safety issues met in science lessons. It is not intended for students to carry out all three exercises, but rather for the teacher to choose the most appropriate exercise for the class. **The risk and dread exercise** is quite difficult to carry out and it is recommended to use it with more able students.

Students may not know the difference between hazard and risk, so this should be pointed out near the start of the lesson.

The risk and dread survey carried out by the Royal Society in 1992 showed that people had more dread of some risks than others.

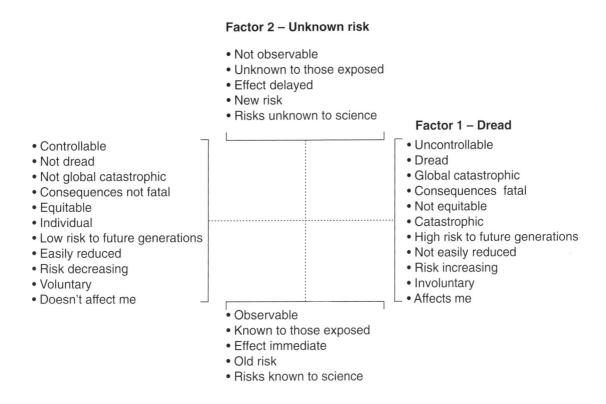

Figure 2 The overall findings of the risk and dread survey
(Source: *Risk, analysis, perception, management,* London: The Royal Society, 1992.)

RS•C

The results of the risk and dread survey show the following:

- Most people are willing to take a risk when they know that they are either in control, the activity is voluntary, the risk is easily reduced or there is a low risk to future generations. Factors such as riding a bicycle, climbing a mountain or using cosmetics come into this category.

- People tend to dread things that are out of their control, activities that are not voluntary and that can effect many people, including future generations. Factors such as handguns, nerve gas and terrorism come in to this category.

The student worksheet **Risk assessments** can be used as a stand alone exercise; rather than using the cartoon you may wish to introduce the exercise in other ways; for example by

- Group discussions

- Scenarios on prompt cards – groups role-play to the rest of the class *eg* someone trips over a bag. The class then have to identify the hazard and decide how to reduce the risk that the hazard poses.

- Accidents that have happened in the past – using the statistics supplied in **Who is safest at school?**

Resources

- Student worksheets
  - Our perceptions of risk
  - The risk and dread exercise
  - Who is safest at school?
  - Risk assessments

Timing

Timing will vary depending on the depth of discussion. The short exercises will take between 30 and 60 minutes. The risk assessment exercise will take a full hour.

Adapting materials

The student worksheets **Our perceptions of risk** and **The risk and dread exercise** can be easily adapted to meet the needs of the less able by removing or rephrasing some of the statements in question 1.

Answers

Our perceptions of risk

1. List in numerical order starting with 'Death as a result of a meteorite falling on you'. Order is 14, 1, 8, 3, 9, 16, 10, 6, 5, 12, 15, 7, 11, 13, 4, 2.

2. Teachers may wish to compare some of these figures when summing up.

The risk and dread exercise

(See also photocopiable sheet with results from the Royal Society study.)

Questions 1–4 Students' own responses.

Students then use the results from the risk and dread survey to answer questions 5–7.

5. They have no control over the situations.

6. They are voluntary activities.

7. They enjoy the activity and willing to take a risk.

Who is safest at school?

1.

Place	Real % of accidents to students
Sports activities	28.9
Gymnasium	18.4
Extracurricular activities	1.0
Corridors, stairs & cloakrooms	13.3
Toilets	1.2
Science laboratories	2.3
D&T workshops	2.2
Food technology rooms	0.4
Classrooms	10.3
Play areas (inside & outside)	13.3
Other	5.3

Table 1 School safety statistics
(Source: Health and Safety Executive Statistics for 1996/97)

2. Science teachers

3. There are strict rules applied during lessons and students are not allowed in laboratories when they are not supervised.

RS•C

Risk assessments

1. The correct order of the statements is as follows:
   – Look for the hazards (3)
   – Decide who might be harmed and how (1)
   – Evaluate the risks (5)
   – Decide what precautions should be taken (4)
   – Record your findings (2)

2. Teachers could use the individual results in making a definitive list.
   The risk assessment for the laboratory will be specific for individual rooms.

RS•C

Our perceptions of risk

1. The following statements list a few of the chances of something happening that we may face in our lives. Sort the statements into order, starting with the statement with the least chance of it happening.

1.	Death as a result of a plane falling on you	1 in 25 million
2.	Death by heart attack	1 in 4
3.	Death by lightning	1 in 10 million
4.	Having asthma as a child	1 in 7
5.	Seriously injuring yourself on exercise equipment	1 in 400
6.	Death as a result of motor cycle racing	1 in 1100
7.	Choosing three balls in the National Lottery	1 in 57
8.	Winning the National Lottery jackpot	1 in 14 million
9.	Being murdered	1 in 100 000
10.	Death as a result of mountain climbing	1 in 1750
11.	Becoming dependent on alcohol	1 in 25
12.	Death related to smoking 10 cigarettes a day	1 in 200
13.	Seeking help for mental illness in your lifetime	1 in 8
14.	Death as a result of a meteorite falling on you	1 in 1 million million
15.	Having a serious fire at home	1 in 160
16.	The next person you meet will be born on the same day and year as you	1 in 25000

2. Discuss with your neighbour any of the statistics that you find surprising or worrying.

The risk and dread exercise

1. Using your own knowledge and experience, complete the following table.
    You must give each item a score, between 1 and 10, for both dread (how much
    you fear something) and unknown risk (you do not know the risks involved).

    A high score indicates that you really dread the item and do not know much about
    the risks associated with it.

    A low score means that you do not fear the item and know about the risks
    associated with it.

Item	Dread	Unknown Risk
Smoking		
Mountain climbing		
Nuclear power		
Sunbathing		
Crime		
Bridges		
Space exploration		
Chemical disinfectants		
Hand guns		
Prescription drugs		
Bicycles		
Terrorism		
Cosmetics		
Alcoholic drinks		
Microwave ovens		
Fireworks		
Radiation therapy		

RS•C

2. When you have finished plot your results on the grid provided. This is the type used by professionals when they are talking about risk

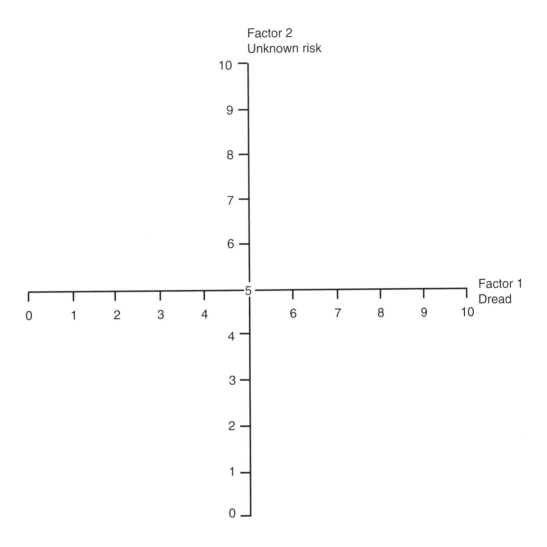

Factor 2
Unknown risk

Factor 1
Dread

3. When you have finished, your teacher will give you the results of a big survey in which people were asked a series of questions on 90 hazards.

4. Compare your results with the ones you have been given. What are the main differences and similarities?

Use the results you have been given to answer the following questions:

5. Why do you think that people are very worried about the risks of terrorism and crime?

6. Why do you think that people are not worried about risks of using cosmetics and sunbathing?

7. The risks associated with mountain climbing and drinking alcohol are well known.
Why do you think that people are not worried about them?

The risk and dread exercise

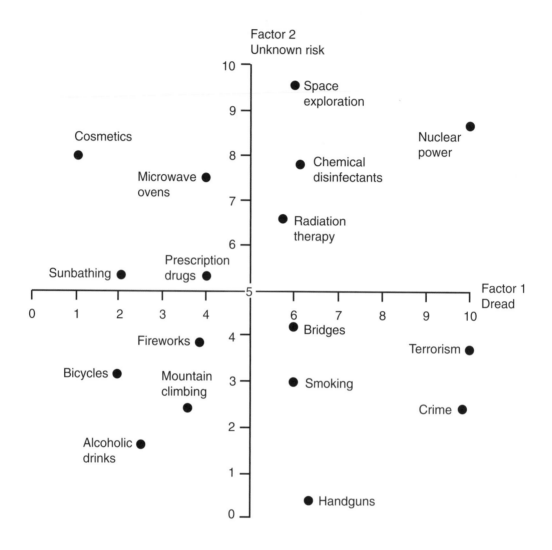

Results from the risk and dread survey
(Source: *Risk, analysis, perception, management*, London: The Royal Society, 1992.)

RS•C

Who is safest at school?

1. Complete the following table by guessing where accidents happen at school.
   Enter your answers in the 'Estimated' column.
   Remember at the end the total should add up to 100%.

Place	Estimated % of accidents to students	Real % of accidents to students
Sports activities		
Gymnasium		
Extracurricular activities		
Corridors, stairs & cloakrooms		
Toilets		
Classrooms		
Science laboratories		
Food technology		
Play areas (inside & out)		
Others		

Now collect the real answers from your teacher so that you can compare them to your estimates.

2. Each year, there are accidents involving staff at school. Study the following bar chart and decide which teachers are safest.

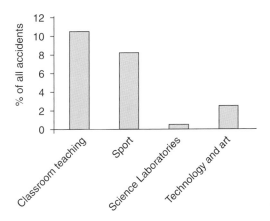

(Source: Health and Safety Executive Statistics for 1996/97.)

3. Why do you think science laboratories are a very safe place to be in school?

Risk assessments

A risk assessment is carried out to make sure that no one gets hurt. It is a careful examination of all the things that could harm you, so that the necessary precautions can be taken.

When you carry out a risk assessment, five things need to be done.

1. Decide who might be harmed and how.

2. Record your findings.

3. Look for the hazards.

4. Decide what precautions should be taken.

5. Evaluate the risks.

Question 1

Put the statements above into the correct order.

Question 2

a) Study Figure 1 on the next page making a note of all the hazards.

b) Carry out a risk assessment of your school laboratory by completing the table. Use your answers to questions 1 and 2 to help.

RS•C

Figure 1 The science class misbehaving

RS•C

Hazard	People at risk	Precaution

c) Using your risk assessment, design a poster containing safety rules for science lessons or write a letter from the 'Head of Science' to 'Parents', explaining why students should always obey safety rules in science.

RS•C

RS•C

Safety symbols

Teachers' notes

Objectives
■ To recognize and understand the meaning of hazard warning symbols.

■ To understand the precautions that should be taken when working with hazardous material to control the risk to themselves and others.

■ To be able to apply this knowledge to everyday situations.

Outline
This activity requires a brief introduction and then the class can work through the student worksheets independently or in groups. Question 4 is intended to be carried out as a homework exercise, but it could be carried out in class if appropriate resources are made available.

Teaching topics

This activity is suitable for 11–14 year olds. However, many teachers may feel that it is appropriate to include it as part of an introduction to chemistry/science for 11–12 year olds. It will fit into any topic that makes use of chemicals.

Teaching tips

This activity can be carried out in groups or done as individual learning

Resources

■ A class set of CLEAPSS student safety sheets or other safety information

■ Everyday items containing hazard symbols (optional)

■ Student worksheets
  – Safety symbols 1
  – Safety symbols 2

Timing

~60 minutes

Adapting resources

The worksheets **Safety symbols 1** and **2** have been included as an example of how adjusting the required level of linguistic skills allows access to a wider range of student ability.

■ Safety symbols 1 has been written for more able students, with a higher reading age.

■ Safety symbols 2 has been written for less able students, with a lower reading age.

RS•C

Answers

Safety symbols 1

1. **Harmful** – These substances are similar to toxic substances but less dangerous.

 Irritant – These substances are not corrosive but can cause reddening or blistering of the skin.

 Highly flammable – These substances catch fire easily.

 Corrosive – These substances attack and destroy living tissues, including eyes and skin.

 Oxidising – These substances provide oxygen, which allows other materials to burn more fiercely.

 Toxic – These substances can cause death. They may be swallowed or breathed in or absorbed through the skin.

 Explosive – These substances will explode when they are set alight.

2. Answers should include precautions such as wearing eye protection, washing hands after use, wearing gloves, using the small amounts and low concentrations, using safety screens, never eat or drink when handling chemicals.

3. Answers depend on what students find.

Safety symbols 2

 Harmful – These materials can make your skin feel itchy.

 Highly flammable – These materials burn easily.

 Irritant – These materials can give you blisters.

 Corrosive – These materials can kill living cells *eg* skin cells.

 Oxidising – These materials give away oxygen, so other materials can burn.

 Toxic – These materials can kill you.

 Explosive – These materials are explosive.

Safety symbols 1

1. Match the following safety symbols and definitions.

harmful or irritant

highly flammable

corrosive

oxidising

toxic

explosive

These substances are similar to toxic substances but less dangerous.

These substances provide oxygen, which allows other materials to burn more fiercely.

These substances catch fire easily.

These substances will explode when they are set alight.

These substances attack and destroy living tissues, including eyes and skin.

These substances can cause death. They may be swallowed or breathed in or absorbed through the skin.

These substances are not corrosive but can cause reddening or blistering of the skin.

2. For each type of hazard, what precautions do you think you should take?
   Eg when using harmful materials always wear safety glasses, avoid contact with hands and do not eat or drink.

3. Safety hazard symbols are used everyday to warn us of danger. See what symbols you can find at home or when you go shopping. Record your results in the following table.

Item	Symbols	Hazard	Precaution
Liquid paper or correction fluid		Highly flammable	Keep away from naked flames

Safety symbols 2

harmful

irritant

highly flammable

corrosive   **explosive**

oxidising

toxic

These materials can give you blisters	These materials give away oxygen so that other materials can burn	These materials can kill living cells, eg skin cells
These materials burn easily		
These materials can kill you	These materials are explosive	These materials can make your skin feel itchy

Meanings

- Cut out the safety symbols, labels and meanings.

- For each symbol find the correct label and meaning.

- Stick them into your exercise books in the right order.

- Try and find a real example of each symbol eg liquid paper has a highly flammable symbol.

The first example has been done for you.

          Harmful   These materials can make your skin feel itchy

PHOTOCOPY P

RS•C

Assessing safety in science experiments

Teachers' notes

Objectives
■ To carry out a risk assessment, prior to doing an experiment.

Outline
This activity includes an example of how students can carry out their own risk assessments and a risk assessment proforma.

Teaching topics

The example experiment and risk assessment is suitable for 11–14 year old students and could be included when teaching about acids and alkalis. Students should understand the concept of an indicator.

The risk assessment proforma is suitable for use by 11–16 year old students, who could use it when planning their own investigations.

Teaching tips

Go through the risk assessment form the first time the class uses it, to make sure that they understand what to put in each column. Ensure the students complete the form with a partner before going on to carry out the experiment.

Resources (per group)

■ 250 cm^3 Beaker

■ Tripod

■ Gauze

■ Bunsen burner

■ Safety glasses

■ Test-tube rack

■ Three test-tubes

■ Red cabbage (3–4 small pieces)

■ Dilute hydrochloric acid (0.5 mol dm^{-3})

■ Calcium hydroxide solution (0.4 mol dm^{-3})

■ Unknown solution A is dilute hydrochloric acid

■ Unknown solution B is calcium hydroxide solution

■ Student worksheet
– Assessing risk in science experiments

Practical tips

Supervise carefully as students boil their cabbage for 10 minutes. Bunsen burners can then be switched off. Beakers can then be lifted carefully off the tripods.

Timing

30 minutes each for the risk assessment and experiment.

Adapting resources

The student worksheet **Assessing risk in science experiments** can be adapted to use with any experiment or investigation by deleting the instructions for making a pH indicator and replacing them with another experiment.

This activity can be made more accessible to the less able student by previously identifying all the hazards on the sheet. This could be done for example by printing key works in bold and using arrows to identify hazardous pieces of equipment.

Answers

A completed risk assessment form

Chemical or microorganism	Procedure or equipment	Hazard	Precaution taken to control risk	Source of information
Dilute acid		Low hazard	Wear safety glasses. Avoid contact with hands	*eg* CLEAPSS student safety sheet
Dilute alkali		Low hazard	Wear safety glasses. Avoid contact with hands	*eg* CLEAPSS student safety sheet
	Boiling the water	Bunsen burner	Turn to yellow flame when not in use. Watch the experiment so that the water does not boil over	My knowledge or teacher
	Lifting hot beaker	Burnt fingers	Leave to cool down a bit or use tongs	My experience

Questions after the experiment

1. Green/yellow

2. Purple/red

3. Red

4. Unknown A is acid and B is alkali.

Assessing risk in science experiments

Whenever investigations are carried out at school you need to decide if the experiment is safe. You should carry out a risk assessment, which is checked by your teacher.

- Read through the method you have been given.

- Identify all the hazards.

- Look up the appropriate student safety sheets.

- Fill in the risk assessment form and ask your teacher to check it.

- Carry out the experiment.

- Answer the questions.

Making a pH indicator

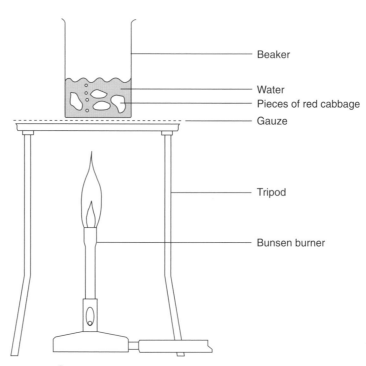

- Boil about 100 cm^3 of tap water in a beaker.

- Add three to four pieces of red cabbage to the boiling water.

- Boil for about 5 minutes. The water should have turned blue or green.

- Turn off the Bunsen burner and allow the beaker to cool for a few minutes.

- Place three test-tubes in a rack. Half fill one with alkali, one with acid and one with deionised water.

- Decant approximately 2–3 cm height of cabbage solution into each test-tube.

- Test solutions A and B with the cabbage solution to see if they are neutral, alkaline or acidic.

Questions

1. What colour is the cabbage indicator when neutral?

2. What colour is the cabbage indicator when alkali?

3. What colour is the cabbage indicator when acid?

4. Are the two unknown solutions A and B neutral, alkaline or acidic?

Risk assessment form

Name of investigation _____

Name of students in group _____

Class _____   Date _____

Chemical or microorganism	Procedure or equipment	Hazard	Precaution taken to control risk	Source of information

RS•C

Risk/benefit analysis

Teachers' notes

Objectives
■ To show that in real life situations risk must be balanced against benefits.

Outline
An example of a risk/benefit balance is given for fire retardants and it is shown how it can be turned into a class exercise.

Teaching topics

This type of exercise is suitable for 14–16 year olds and could be used when teaching about industrial chemical processes such as the manufacture of chlorine, sulfuric acid, iron and steel, aluminium, polymers or fertilisers. This concept could also be applied when considering the Principles of Green Chemistry as outlined in the RSC publication Green Chemistry.[1]

Background information

The objective of risk/benefit analysis is to identify risks from the procedure/chemicals and balance then against the overall benefits of the finished products. It is a framework for determining how far it is worth controlling the production, use, storage and disposal of existing substances in order to achieve reductions in the risk to human health and the environment. Cost is a key player in the analysis and if it is relatively cheap to add an extra control to reduce the risk, then the control should be added.

Risk analysis does have its limitations and if taken to its extreme many of the everyday things that we take for granted could be banned. Also, risk analysis on its own does not take into account adjustments needed as the economic situation, technology or required resources change. Therefore more often or not a compromise has to be made. In practice risks are weighed up against benefits, until finally a decision is made. If the benefits outweigh the risks then it would seem reasonable to go ahead with production, however if the balance tips the other way then production would have to be questioned.

This type of analysis is frequently carried out in industry.

RS•C

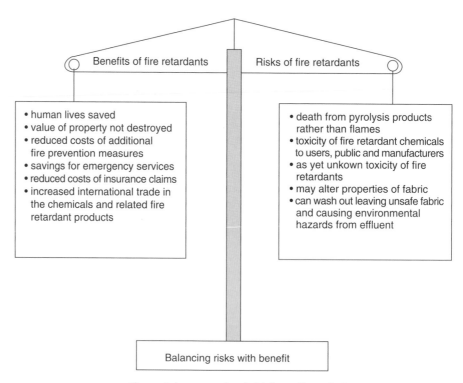

Figure 3 An example of risk/benefit analysis
(Reproduced with permission from *Green Chlorine*, York:
Chemical Industry Education Centre, 1997.)

Sources of information

A section on risk / benefit analysis is included in the teachers' notes of *Green Chlorine*, York: Chemical Industry Education Centre, 1997.

Teaching tips

This activity could be carried out after learning about some aspect of industrial chemistry. In groups the students should discuss and make a list of the risks posed by manufacturing and the benefits gained from the product. Effects on the environment, local and global economy and general issues of safety should be included. A whole class brainstorming session could be used to get started.

Once the benefits and risks have been identified the risk benefit analysis student sheet can be completed.

Resources

■ Scissors

■ Glue

■ Student worksheet
– Risk/benefit analysis

Timing

30 minutes

Risk/benefit analysis

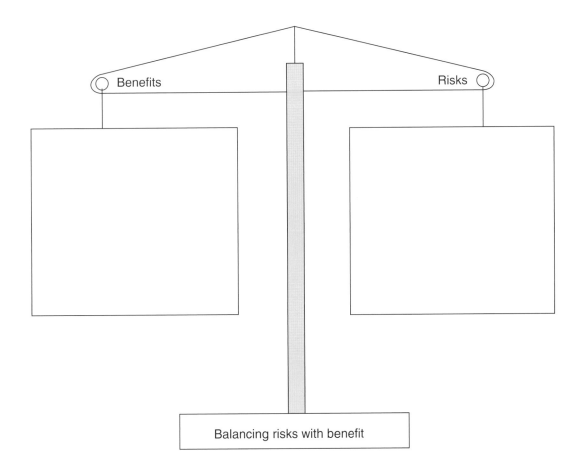

- List all the benefits in the box on the left hand side of the balance.
- List all the risks in the box on the right hand side of the balance.
- Cut out the pieces of the balance.
- Stick the balance in your book, showing which way it is tipped.
- Based on your balance, do you think the product should be made?

RS•C

RS•C

Alice Hamilton – safety, hazards and risk

Teachers' notes

Objectives

■ To show how scientific work can be affected by the contexts in which it takes place.

■ To learn about the powers and limitations of science in addressing industrial, social and environmental questions.

■ To show the importance of carefully planning an investigation.

■ To understand the importance of controlling hazards and reducing risks to maintain a safe working environment.

Outline

The introductory material about Alice Hamilton illustrates that scientific evidence can lead to changes in legislation and improved working conditions. The resource is presented as three case studies, followed by suggestions and guidance for carrying out project work.

■ Case study A The typhoid problem at the start of the century

■ Case study B Alice Hamilton investigates industrial lead poisoning

■ Case study C Alice Hamilton investigates carbon monoxide poisoning

■ Project work.

Teaching topics

These activities are suitable for 14–16 year olds and could be included when teaching about useful products from metal ores and rocks or about carbon monoxide and different types of reactions.

Background information

The life and work of Alice Hamilton is used to illustrate the importance of safety when working with and dealing with chemicals. At the turn of the century, the Europeans were more aware of industrial poisoning than the Americans. In the USA there was no legal health protection at work or compensation for 'poisoned' industrial workers. The situation in Britain was much better. By 1928 the Home Office was making a new set of rules for the control of lead poisoning in storage battery plants. Unlike in the United States of America, the trade unions' and factory experts' representatives in Britain had equal rights and so it was easier for legislation to be made that looked after the interests of the factory workers to be written and subsequently enforced. Alice Hamilton was born in and lived in the United States of America. She used her medical background and scientific training to fight the American authorities. Her many investigations led to safer working conditions for thousands of people and, in many cases, legislation concerning aspects of public health. Her impact on the industrial society was enormous. Her reputation went before her to the extent that some managers tried to cover up the poor working conditions and told sick workers to stay at

RS•C

home during her visits. Needless to say, they were found out because Alice made it her business to talk to the ordinary people in the towns and villages.

Alice Hamilton carried out many interesting studies in her lifetime, covering a wide range of chemicals. This section focuses on the typhoid problem, lead poisoning and carbon monoxide poisoning, as they each illustrate different aspects of scientific enquiry, and will easily slot into a teaching scheme.

Health – long term effects

The project work approach could be adapted to other people and other safety problems eg Marie Curie and radiochemistry, comparing working conditions now and them. A good resource and starting point is *100 years of Radium, Marie Curie and the History of Radiochemistry*, Hatfield: The Association for Science Education and The British Society for the History of Science, 1999 (ISBN 0 86357 299 5).

Source of information

The autobiography of Alice Hamilton provides details of many of her other investigations.

A. Hamilton *Exploring the Dangerous Trades*, North Eastern University Press, 1989 (reprint, now out of print).

http://www.distinguishedwomen.com/biographies/hamilton-a.html (accessed June 2001)

Teaching tips

Alice Hamilton investigates public health

This could be introduced by explaining that, when Alice Hamilton lived, not much was known about the hazardous nature of many chemicals or the risks they posed. The materials worked with were slowly poisoning many of the factory workers. Over the years Alice Hamilton played a prominent role in bringing these concerns to the attention of the government and industry.

The resource can be used to emphasise the importance of observing safety rules today, such as no food or drink to be taken into the laboratory and washing hands after handling chemicals etc. This could then lead into a discussion on the importance of reducing the risks we face in the workplace, and an introduction to risk assessment.

Case study A

This can be used as a model for carrying out an investigation or used to focus investigational skills such as observation and testing out different theories.

Case study B

This could be used prior to carrying out a smelting experiment in class or as a follow up piece of work.

Refer also to *Classic Chemistry Experiments*, Royal Society of Chemistry: London, 2000, experiment number 35.

Case study C

This activity helps students to become familiar with the information contained on the CLEAPSS students safety sheets.

RS•C

Project work

Suggested project questions are found at the end of each case study. When carrying out project work, the student's interest could be captured by focusing on a current or local issue.

Suggestions and guidance for carrying out project work
This is an example of how a project could follow on from Case Study C on carbon monoxide poisoning. This approach could be applied to almost any topical issue.

Introduce the project as a real problem, needing real solutions and answers.

Newspaper headlines You may wish to use a couple of headlines to stimulate interest and discussion at the start of the project. These articles could be transferred on to an OHT slide.

Context The students could be acting as reporters for newspapers, magazines, local councils etc. As a reporter it is their job to investigate the problem, looking for evidence to support or reject an initial claim, and suggesting ways to try and minimize the problem and public fears. It is important to emphasise and keep the project focused on science. This can be achieved by using prompt questions at the start, or by stating expected outcomes. We all know how easy it is to be sidetracked and go off at a tangent. Some type of final presentation either written or in the form of a poster or as an oral presentation or debate should be the final outcome. For further details see Teaching students to communicate ideas in science, p. ix.

The material is suitable for either individual or group work, depending on time constraints and student ability. If working in groups, the group should be no larger than 4, with each group member having a specific task to carry out, (eg each group member taking a different point) so that real team work is achieved and not all the work is done by one or two dominant characters.

Resources

- CLEAPSS or SSERC student safety sheets or HAZcards

- Encyclopaedia

- Science dictionary

- Internet access

- Overhead projector (OHP)

- Student worksheets
  - Alice Hamilton investigates public health
  - Carbon monoxide: introducing the problem
  - Carbon monoxide project work
  - **A detector calls** three page reprint article from Chemistry in Britain. (Reproduced with kind permission from Chemistry in Britain R. Kingston *Chemistry in Britain*, 1999, **35(4)** 44–46.)

Timing

Case studies
Each study will take about 60 minutes

RS•C

Project work

As with any type of project work the timing is very important and tight deadlines must be set and enforced. For example, students should be told in advance when the resource room/ICT room is available to use during lesson time, when their initial research should be completed by, when poster making materials etc will be available, and of course a completion deadline. Experience has shown that if deadlines are not enforced the project can drag on for a very long time, because in this type of work there is always room for improvement as new ideas come along later.

At the onset of the project it is important to decide how much lesson time and homework time you wish to spend on the project, and pass this information on to the class.

Opportunities for using ICT

The project work approach presents opportunities for using:

- The Internet

- CD-ROMs

- Word processing

- Software packages suitable for presenting work.

Opportunities for key skills

- Communication

- Application of number

- Working together.

Answers

Alice Hamilton investigates public health

1. 101

2. She spent her life working with dangerous and poisonous chemicals. She also lived with people who often contracted fatal contagious diseases.

3. She wanted to help the poor people who were living and working in poor conditions. She believed that everyone had the right to be safe at work and should not have to suffer because employers did not want to invest in cleaning up the work place.

4. Most of these chemicals have been superseded by safer chemicals today. However, you will still find that some of the chemicals are in use today in more controlled applications.

5. People are more aware of safety today and employees in the workplace are protected by the law. There are many documented studies surveys on the safety and use of chemicals today. When new chemicals are produced, the safety of that material is studied.

Case Study A

1. Flies were feeding on typhoid infected waste from overflowing privies and household waste. The flies landed on food and milk, which was consumed by the people who were now infected.

RS•C

2. She collected flies from the privies and kitchens. She dropped them into broth and incubated the tubes for different amounts of time.

3. The results showed that the typhoid bacteria were present, thus supporting Hamilton's theory.

4. To cover up food and drink. To wash your hands before cooking, eating or drinking and after going to the toilet. To wash down the kitchen surfaces before preparing food and to clear up afterwards.

5. At the local pumping station on West Harrison Street, a break occurred which resulted in an escape of sewage into the water pipes. The public drank the infected water for three days.

6. When the results from Hamilton's theory supported her theory she thought that she had found the cause of the epidemic and so stopped her investigation.

Case Study B
1. Lead metal – low hazard. Lead oxides are toxic. Harmful if swallowed or the dust is breathed in. The compound builds up over a long period of time. It may harm an unborn child.

2. To avoid breathing in dust or fumes, wear a facemask. Make sure that protective clothing is worn while working with the compounds and that the clothing is removed before going home. Always wash your hands after handling lead compounds.

3. Hamilton could have made the following recommendations to:
   a) the enamel industry – to clean up the lead dust, to provide a common room for the employees to eat their lunch, to ensure that the employee changed their clothes and washed their hands as they left the factory floor. A clean room being provided for changing and washing.

   b) the smelting industry – as for the enamel industry and to regularly service smelting machinery and improve the ventilation so that the lead fumes could escape.

4. The painters were proud people and wanted their work to be of the highest quality. The lead and turpentine paints gave better results than other paints. If they used cheap paints, with poor results then perhaps no one would employ them. They were prepared to risk their health but not their reputation.

Case Study C
1. CO

2. $2C + O_2 \rightarrow 2CO$

3. Toxic and flammable

4. CO is a colourless, odourless, tasteless and non-irritant gas and there were no chemical tests available.

5. Poor ventilation when using gas fires.

6. Good ventilation is needed.

RS•C

Alice Hamilton investigates public health

Alice Hamilton (1869–1970)
(Reproduced with permission from Bentley Historical Library, University of Michigan.)

Alice Hamilton went to a girls' boarding school where very little science was taught. Her ambition was to be a doctor, so she had to learn chemistry and physics in the summer holidays. She was successful and went to Michigan Medical School.

In 1897 Alice decided to live at Hull House settlement home, in Chicago. Settlement homes offered help to poor people such as immigrants from run-down inner cities. She was concerned about the dirty living conditions and working conditions in the factories. She investigated typhoid fever and tuberculosis.

She spent thirty years of her life investigating factory working conditions and studied the effect of dangerous chemicals on humans. As a result of her work, in 1937 new laws were passed in the USA giving compensation for industrial diseases.

Alice Hamilton's investigations into hazardous materials

She:

■ Studied white lead and lead oxide substances that were commonly used as pigments in the paint industry and recommended safer working conditions.

■ Investigated the poisonous effects on workers of manufacturing explosives.

■ Studied aniline dyes, carbon monoxide, mercury, tetraethyl lead, radium (in wristwatch dials), benzene, the chemicals in storage batteries, carbon disulfide and hydrogen sulfide (created in the manufacture of viscose rayon).

Questions

1. How old was Alice Hamilton when she died?

2. Why might you be surprised by her age?

3. What do you think motivated Alice Hamilton to carry out her investigations into hazardous materials?

4. Try and find out which of the chemicals she studied are in use today? Hint: You could look up the chemical names on a science CD-ROM or use an encyclopaedia.

5. Does this tell us anything about safety today?

Case Study A The typhoid problem at the turn of the century

Typhoid is a bacterial disease that causes a high fever and attacks the intestines. Bacteria live in warm, wet conditions where there is a good food supply.

Background information

■ **Drinking water** – was taken straight from the lake with no chlorine treatment.

■ **Precaution taken against pollution** – daily water samples were taken to make cultures. The next day the results were published, advising whether or not to boil the water.

■ **Assumption** – housewives would look at the results and act upon the advice.

Alice Hamilton Investigates the 1902 Typhoid Epidemic at Hull House in Chicago

Facts:

■ Hull house was at the centre of the epidemic.

■ Hull house used the same main water and milk supply as less affected areas.

Deduction:

■ It must be a local problem.

Investigation

■ **Observations** – around the local streets showed that some outside privies (toilets) were overflowing into the backyards and streets, mixing with the rainwater. The whole area was very dirty, the plumbing was out of order and there were flies everywhere.

■ **Knowledge** – during the Spanish-American war in 1898, studies had shown that typhoid was spread by house flies.

■ **Hypothesis** – the flies were feeding on typhoid infected waste and then landing on the food and milk.

■ **Experiment** – Alice collected flies from privies and kitchens. She dropped them into broth and incubated the tubes for different amounts of time. The results showed 'typhoid bacteria'.

■ **Conclusion** – dirty living conditions was the cause of the typhoid epidemic.

■ **Outcome** – a public enquiry resulted in a complete reorganization of the Public Health System. There were regular tenant house inspections.

Board of Health kept the real cause to themselves

At the local pumping station on West Harrison Street, a break had occurred which resulted in an escape of sewage into the water pipes and for three days the neighborhood drank the water, before the leak was discovered and stopped.

Questions

1. What did Alice Hamilton believe was the cause of the 1902 typhoid epidemic?

2. How did she test out her theory?

3. Did the results support or undermine her theory?

4. As a safety officer what advice would you give to the local people to try and avoid typhoid in the future?

5. What was the real reason for the 1902 typhoid epidemic?

6. Why do you think Alice Hamilton did not test the local water supply and find the real cause sooner?

Project or extension work

Find out about the measures that are taken today to ensure that typhoid epidemics are no longer common in this country.

Getting started – carry out an Internet search on typhoid.

Case Study B Alice Hamilton investigates industrial lead poisoning

The enamel industry – 1912

When the enamel workers were on strike in 1912, Alice Hamilton took the opportunity to examine them. She was looking for the 'lead line', which is a deposit of black lead sulfide in the cells of the lining of the mouth. It is usually clearest on the gum along the margin of the front teeth. The line is formed when lead reacts with protein in food that is being eaten and indicates lead poisoning. The results were alarming, and showed 54 out of 148 workers had 'lead lines'. Examining hospital and doctor's records confirmed the extent of severe lead poisoning.

Walking around the factories Alice Hamilton observed a lot of lead dust and men eating their sandwiches in the same rooms without washing their hands or changing out of their work clothes.

Lead smelting

A similar study revealed lead poisoning was just as widespread in the smelting industry. The main dangers came from:

- **Dust** when the ore was ground, when the charge was fed into the furnaces and when the flues were cleaned out.

- **Lead fumes** that escaped from the furnace.

Once again the workplace was very dusty and hands were not washed when food was eaten. Machinery was simple and not kept in good order, so lead fumes escaped into the very hot rooms which were not well ventilated.

Painting

Around 1913, there were two hazards associated with the painter's trade; lead and turpentine. The risks they posed were well known. However, the paint that gave the best results was lead and turpentine based. The newer cheaper paints were based on a leadless pigment and naphtha and was not liked. The painters preferred the lead based paint and were willing to take the risk of lead poisoning. They often complained about the headaches and nausea caused by turpentine fumes.

As Alice Hamilton investigated she was sure that once again the 'unwashed hand theory' would be supported. The investigation showed that the work place was very dusty, especially when surfaces were being rubbed down and white and red lead were being mixed with oil. Hospital records confirmed a number of cases of lead poisoning amongst painters.

By the 1940s iron and titanium oxides replaced the lead oxides used in paints.

Questions

1. Using the appropriate student safety sheet, what is the main hazard posed by lead and lead oxide?

2. What precautions should be taken when working with lead compounds?

3. When Alice Hamilton wrote her report, what general recommendations do you think she made to:
   a) the enamel industry?
   b) the smelting industry?

4. Why do you think painters were willing to take the risk of using leaded paints when there were alternatives available?

Extension or project work

Find out more about the effects of lead poisoning and then decide whether you would you be willing to use a lead based paint every working day.

Case Study C Alice Hamilton investigates carbon monoxide poisoning

In 1919, carbon monoxide poisoning was widespread, especially among coal miners and garage workers.

Sources of carbon monoxide in mines

- Dynamite was used for blasting. A dynamite blast produces a gas, which can be up to 34% carbon monoxide.
- Slow combustion of coal veins, in a limited oxygen supply.
- Coal dust explosions can produce enough carbon monoxide to kill miners.

Sources of carbon monoxide in garages

- Exhaust fumes could have up to 12% carbon monoxide.

Properties of carbon monoxide

- Colourless gas
- Odourless gas
- Tasteless gas
- Non-irritant.

Symptoms of carbon monoxide poisoning

- Headaches
- Weakness especially in the legs
- Dulled mind
- Loss of consciousness leading to brain damage.

Method of investigation

- **Observation of the work place.** This could not determine the amount of carbon monoxide, but it could be used to monitor the level of ventilation.
- **Results of carbon monoxide poisoning.** Where possible, this study used victims of carbon monoxide poisoning to investigate the long term effects.

Results

Unfortunately the results were inconclusive and the study could not continue. Chemical tests for detecting carbon monoxide were unavailable at the time and so it was not possible to detect carbon monoxide levels in garages and traffic tunnels etc.

Use the information above and the student safety sheet or HAZcard on carbon monoxide to answer the questions.

Questions

1. What is the chemical formula for carbon monoxide?
2. Write an equation for the slow combustion of coal in a limited oxygen supply.
3. Which safety symbol(s) would you assign to carbon monoxide?
4. Why do you think it was difficult for Alice Hamilton to investigate carbon monoxide poisoning?
5. What is the main cause of carbon monoxide poisoning in the home today?
6. What precautions should be taken to avoid carbon monoxide poisoning today?

Extension or project work

Carbon monoxide the silent killer – Find out how carbon monoxide levels are detected and monitored today (at home and in industry.)
Find out how carbon monoxide slowly starves the brain of oxygen.

Carbon monoxide: introducing the problem

Carbon monoxide the silent killer – claims about 200 victims a year in the 1990s according to the Royal Society for the Prevention of Accidents (RoSPA).

02 February 1999,
The Times, p.2
Home news – News in brief

Family die of fume poisoning

A family of four were found dead at their home yesterday from carbon monoxide poisoning. Neighbours of Jeffrey Cheetham, 37, his wife, Beverley, 36, and their sons Christopher, 10, and Carl, 8, called police after noticing that curtains at the house in Brimington, Derbyshire, had been drawn since Sunday. After officers broke in, several were overcome by gas and had to be taken to hospital for tests.

14 March 1999,
The Sunday Times
Features – Education

How to be safe but sorry

By Jennie Brist

As for awareness:
a university's obsession with promoting personal safety advice hits you as soon as you arrive at your freshers' fair. Bombarded with leaflets about the danger of drink, drugs, sex, meningitis and carbon monoxide poisoning,

Carbon dioxide (CO$_2$) and monoxide (CO)

Substance	Hazard	Comment
Carbon dioxide gas	DANGER	Can cause asphyxiation if proportion of carbon dioxide in the air becomes too high, eg as a result of the rapid evaporation of the solid in a confined space, or, in some African lakes, released from decaying organic matter. As it is denser than air, may build up in low areas, eg in caves. For a 15-minute exposure, concentration should not exceed 27,400 mg m^{-3}. About 0.04% present in normal air as compared with about 0.03% 50 years ago. This increase is the result of burning fuels in motor vehicles, power stations, etc. This is in turn believed to be causing a very gradual rise in the temperature of the Earth (global warming) as a result of the greenhouse effect.
Carbon dioxide solid "dry ice"	DANGER	Causes frostbite (burns) – needs careful handling. If it evaporates rapidly in a closed vessel may cause explosion or, in a confined space, may cause asphysxiation as the air is forced out.
Carbon monoxide gas	TOXIC EXTREMELY FLAMMABLE	Toxic if breathed in, with the danger of serious damage to health by prolonged exposure. May cause harm to the unborn child. As little as 0.01% can cause headaches. The gas has no taste or smell. Often formed when hydrocarbon fuels burn in a limited supply of air, eg car engines especially in confined spaces, or gas-powered water heaters with poor ventilation. Every year causes many deaths in the home. Traces also occur in cigarette smoke and are implicated in heart and artery disease. For 15-minute exposure, concentration should not exceed 349 mg m^{-3}. Forms explosive mixtures with air and oxygen. Mixtures with air between 12% and 74% carbon monoxide by volume are explosive.

© CLEAPSS 1997

Carbon monoxide project work

You are a reporter for the magazine 'Living in a scientific world'.

Your job is to investigate whether carbon monoxide really is a killer and what can be done to stop this deadly killer. The findings of the investigation should be written up for the magazine. Your editor will allow this feature article two sides of A4, she will expect it to be clearly illustrated with pictures, diagrams, charts and tables.

Your article should include the following:

- Evidence that carbon monoxide still kills in the late 1990s.

- The physical properties of carbon monoxide that make it a silent killer.

- Sources of carbon monoxide and the risk it poses in the home.

- The symptoms and science of carbon monoxide poisoning.

- Methods of detecting and alerting people to the presence of carbon monoxide.

- Recommendations for protecting your family from carbon monoxide poisoning.

Getting started

You will need to carry out some research. A good way to start is to work through some of the points listed below.

- A simple search of 'Carbon Monoxide Poisoning', on a CD-ROM such as 'The Times' newspaper, should bring up several up-to-date articles.

- Visit the following websites to find out about the symptoms of carbon monoxide poisoning, common questions and answers about carbon monoxide and the latest in carbon monoxide detection technology:

  http://www.howstuffworks.com/smoke.htm (accessed June 2001)

  http://www.rospa.co.uk/ (accessed June 2001)

  http://www.qginc.com (accessed June 2001)

  http://www.freenet.msp.mn.us/people/guest/pubed/cofaq.html (accessed June 2001)

- Surf the web by doing a keyword search.

- Ask your teacher for a copy of the article, written by Rob Kingston, from Chemistry in Britain called 'A detector calls'.

- Use a word processor, drawing or desktop publishing package to produce the final article.

A **detector** calls

Smoke inhalation and carbon monoxide poisoning claim hundreds of lives every year. **Rob Kingston** looks at the detection technologies that can cut the death toll

SMOKE ALARMS SAVE LIVES. IT'S A FAMILiar message, and one that is borne out by statistics. In 1997, there were over 70 000 house fires in the UK, resulting in around 550 deaths. In those cases where a smoke alarm was present in the area of the fire, the death rate was around two deaths per 1000 fires; where there was no alarm this figure was nine per 1000. The overall rate of almost eight fatalities per 1000 fires reflects the sad fact that in only 30 per cent of cases was an alarm present in the area of the fire.

Fires in the home can spread with astonishing speed, releasing asphyxiating quantities of smoke within minutes – and ironically, modern flame retardant materials may actually increase the production of toxic fumes (see *Chem. Br.*, June 1998, p 21 for a discussion of flame retardant technology). Smoke detectors, which are widely available for less than £10, can give an ear-piercing warning while a fire is still only smouldering, providing valuable extra minutes that can mean the difference between life and death for the occupants of a burning house.

Sniffing out smoke

But what actually goes on inside that little round white box with its reassuring flashing light? All smoke alarms consist of two basic components: a sensor to detect smoke particles in the air and a siren to give the warning.

There are two common types of sensor in use in domestic smoke alarms: photoelectric and ionisation. The first of these simply uses light to 'see' the smoke. Inside the detector is a light source, with a photoelectric cell – which produces an electric current when light is shone on it – set back at 90 degrees to the light beam. When no smoke is present, the light beam passes straight across the detector and misses the photoelectric cell, so no current is produced. In the event of a fire, however, smoke particles enter the detector and scatter the light, reflecting some of it onto the cell (*see Fig 1*), and so producing a small current.

Once the amount of light falling onto the cell – and thus the current produced – reaches a set level then the siren is activated.

The second, and more common, type sensor, the ionisation detector, contains a small amount (typically less than 0.2 m americium-241 (*see Box 1*), as the ($^{241}AmO_2$). Americium-241 emits both particles and low energy gamma rays, bu the α-particles that are vital to the operati the detector.

The ^{241}Am is situated in the ionis chamber of the detector. This consists o charged metal plates, one of which has a to allow the α-particles emitted by the rad tive source to pass through (*Fig 2*). The o ticles ($^4He^{2+}$) collide with molecules of ox

1. Americium

Americium (Am), with atomic number 95, was the fourth artificial transuranium eleme to be identified. It was discovered by Glenn Seaborg and others at the University of Chi go, US, in 1944. The most stable isotope is ^{243}Am, with a half-life of over 7500 years, b it is ^{241}Am, with a half-life of 432 years, that is used in smoke alarms. Americium-241 produced in nuclear reactors as a beta-decay product of plutonium-241, which itself formed ultimately from uranium-238:

$$^{238}_{92}U \xrightarrow{+^1_0n} {}^{239}_{92}U \xrightarrow{\beta\text{-decay}} {}^{239}_{93}Np \xrightarrow{\beta\text{-decay}} {}^{239}_{94}Pu \xrightarrow{+^1_0n} {}^{240}_{94}Pu \xrightarrow{+^1_0n} {}^{241}_{94}Pu \xrightarrow{\beta\text{-decay}} {}^{241}_{95}Am$$

Americium oxide costs around $1500 a gram, a price that has remained virtually u changed since it was first offered for sale by the US Atomic Energy Commission in 196 The low cost of household smoke detectors is made possible because one gram of Am provides enough active material for over 5000 detectors.

Fig 1. Photoelectric smoke sensor

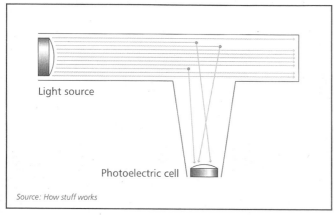

Light source

Photoelectric cell

Source: How stuff works

Fig 2. Ionisation chamber of ionising smoke alarm

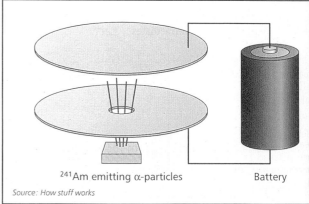

^{241}Am emitting α-particles

Battery

Source: How stuff works

and nitrogen in the air between the plates, ionising them. A small electric current therefore flows continually between the charged plates, as long as the air is clear. If smoke particles, which are also highly ionised, enter the chamber they attach to the oxygen and nitrogen ions, neutralising them and disrupting the current. The control circuits of the detector sense this drop in current and activate the alarm.

It is difficult to say which is the better type of smoke alarm, because there are arguments for and against both types of sensor. Photoelectric detectors tend to be more expensive and less sensitive; however, they are better at detecting the large smoke particles that are produced by smouldering fires, and are less likely to give false alarms due to dust, or grease or steam from cooking. Ionisation detectors respond better to invisibly-small particles, such as those produced by hot, blazing fires. Some authorities recommend installing at least one of each type to be sure that a fire will be detected.

The silent killer

Fire is a very visible hazard. But a much more sinister killer, carbon monoxide, claims as many as 200 victims a year, according to Royal Society for the Prevention of Accidents (RoSPA) estimates. Earlier this year, the death of a family of four, who were overcome by the gas in their Derbyshire home, made headlines in the national press, but many other incidents go unreported.

CO is produced during incomplete combustion of fuels such as natural gas, oil, coal and wood. If it is breathed in, it enters the bloodstream where it binds to haemoglobin (Hb), forming carboxyhaemoglobin (COHb). Because CO binds much more strongly than O_2 to haemoglobin, it effectively 'ties up' the Hb and reduces the oxygen-carrying capacity of the blood. If enough Hb is converted to

COHb, the body becomes starved of oxygen, leading eventually to unconsciousness and death (*see Table 1*). And because COHb has a long half-life in the body – about five hours – even low concentrations of CO can, over time, lead to fatal levels of COHb in the blood.

The most frightening aspect of carbon monoxide is that it is completely undetectable by human senses: it is odourless, colourless and tasteless. Furthermore, the initial effects of CO poisoning (headache, dizziness, nausea) often go unheeded because they are easily confused with the symptoms of common illnesses such as influenza.

However, detectors are available to alert people to the presence of CO in their homes, using a number of different detection methods. The first commercial sensor was the so-called Taguchi sensor (named after its inventor), which was developed by Figaro Engineering of Osaka, Japan, in the 1960s. The Taguchi sensor is an example of a metal oxide semiconductor (MOS) sensor (*see Box 2*), and contains a heated tin oxide pellet, the conductivity of which varies according to the concentration of CO present, allowing a crude measure of ambient CO levels. MOS sensors are not ideal for home use, however: they are not specific to CO and can give false alarms with a number of other substances such as hair spray, air freshener or paint fumes. Another problem is that due to their high power requirement they must be mains-powered: an

important consideration, because many CO poisoning incidents occur when appliances such as gas lamps and kerosene heaters are used during power cuts or in areas without an electricity supply. The need for mains power also dictates the positioning of such detectors in the home, because the intended location may not have a power socket nearby.

Despite these limitations, however, many of the CO detectors on sale today still use MOS technology.

Improved response

The first battery-powered CO detectors were made possible by using an optical detection technique based on colour chemistry. These so-called biomimetic sensors use a similar colour change to that which occurs in the formation of COHb in the blood, and are therefore more specific to carbon monoxide. However, because this type of sensor continually absorbs even very low levels of CO that are always present in the air, it is difficult to obtain a reliable baseline from which to measure the concentration of gas present. Biomimetic sensor technology is still the subject of much research, however. Quantum Group, an environmental health company based in San Diego, US, was the first company to produce commercial biotechnology-based CO sensors. Quantum's detectors use 'artificial haemoglobin', produced with the aid of genetically engineered organisms. These organisms produce molecular structures that resemble a 'keyhole', the 'key' to which is the CO molecule. The presence of CO is detected by measuring the

2. MOS sensors

The gas sensing element in a metal oxide semiconductor (MOS) sensor consists of two coiled-wire heating elements, separated by insulating material and embedded into a block of metal oxide semiconductor. The semiconductor used (typically tin oxide) is an *n*-type, in which electric current is predominantly carried by the flow of free electrons, as opposed to a *p*-type, in which the major carriers are positively-charged 'holes' in the conduction band.

Because the number of free carriers is a function of the surface temperature of the semiconductor, the heating elements are used to maintain a temperature of about 250°C. The high temperature also prevents water condensation forming on the surface of the semiconductor, which would reduce conductivity.

Usually, the semiconductor surface has oxygen atoms from the air chemisorbed onto it. When an oxidisable (*ie* reducing) gas such as CO is present, it reacts with the trapped oxygen, causing free electrons to be released into the semiconductor's conduction band and so increasing its surface conductivity. Note that any reducing gas will have this effect, so MOS sensors are not specific to carbon monoxide. □

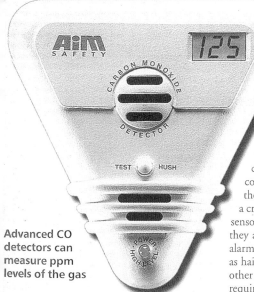

Advanced CO detectors can measure ppm levels of the gas

3. A quick check

Another type of CO detector is the EI200 CO Spot, produced by E. I. Electronics of Shannon, Ireland. This is a small self-adhesive plastic badge, which can be fixed to a wall near a potential source of carbon monoxide. On the badge is a small area of orange crystals, which turn black on exposure to carbon monoxide. Even a small colour change can be noted by comparing the colour of the crystals to the orange surround.

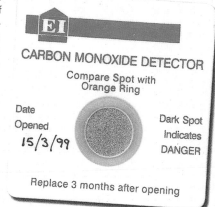

The chemistry involved is based on the Wacker reaction. The active part of the detector comprises a palladium(II) chloride dihydrate–copper(II) chloride co-catalyst system, which acts in a pseudo-homogeneous manner. The colour change is due to the palladium(II) being reduced to palladium(0), *ie* Pd metal.

The reaction is reversible; the key steps being set out below:

Reduction (occurs in the presence of CO): the carbon monoxide reacts with the palladium(II) chloride dihydrate, forming carbon dioxide and reducing Pd(II) to Pd(0):

$$CO + Pd^{II}Cl_2 \cdot 2H_2O \longrightarrow CO_2 + Pd^0 + 2HCl + H_2O$$

Oxidative regeneration (occurs when the sensor is exposed to air which is free from CO): firstly, the palladium metal is oxidised back to Pd(II) by the copper(II) chloride (which is itself reduced to copper(I) chloride):

$$Pd^0 + 2Cu^{II}Cl_2 \cdot 2H_2O \longrightarrow Pd^{II}Cl_2 \cdot 2H_2O + Cu^I_2Cl_2$$

then the copper(I) chloride is oxidised back to copper(II) chloride by atmospheric oxygen:

$$Cu^I_2Cl_2 + 2HCl + H_2O + \tfrac{1}{2}O_2 \longrightarrow 2Cu^{II}Cl_2 \cdot 2H_2O$$

As can be seen from the above reactions, the presence and retention of both water and hydrogen chloride within the sensor is vital to allow proper, reversible functioning of the device. This is achieved by supporting the active crystals on hydrophilic silica gel, incorporating hygroscopic, chloride-containing materials such as calcium chloride, and adding strong acids such as silicomolybdic acid ($H_8[Si(Mo_2O_7)_6] \cdot 28H_2O$) to the system.

While these detectors are very cheap, they do have their limitations. As with MOS sensors, gases other than CO such as ammonia and hairspray fumes can cause false alarms, and the crystals gradually darken on exposure to the air even in the absence of CO. For this reason the badges have to be replaced every three months. However, they do provide a useful, and affordable, 'quick check' for the presence of carbon monoxide. □

---

transmittance of light through the keyhole, using a light emitting diode (LED) source and a photodiode detector.

The most recently introduced type of detectors, which are claimed to overcome the problems of other types, are those that make use of electrochemical technology. With this type of sensor, any CO that is present diffuses into the detector where it reacts with oxygen in the air, according to the general equation:

Sensing electrode:
$$CO + H_2O \rightarrow CO_2 + 2H^+ + 2e^-$$

Counter electrode:
$$\tfrac{1}{2}O_2 + 2H^+ + 2e^- \rightarrow H_2O$$

Overall: $CO + \tfrac{1}{2}O_2 \rightarrow CO_2$

The current flowing between the two electrodes (through an external circuit) is proportional to the amount of CO present over a wide concentration range. This makes it possible for the detector to differentiate between acute, high concentrations and lower, but still hazardous levels, and to give warnings accordingly. To avoid a build up of carbon dioxide in the sensor, the electrode reactions take place under acidic conditions. Under these conditions the process needs a catalyst; platinum is usually chosen because it can form a range of

chemisorbed surface species and so lower the activation energy of the intermolecular reactions. As a result, the response time of the electrode process is fast – it is measured in seconds.

Setting the standard

Among the most vociferous proponents of electrochemical detection technology is George Kerr of Aim Safe-Air Products, a manufacturer of CO detectors based in Austin, Texas, US. Kerr points out that even approved optical- and MOS-based detectors, which meet the required safety standards, often fail to operate correctly, either giving false alarms or – more seriously – failing to sound the alarm when dangerous concentrations of CO are present. Kerr's view is backed up by the Chicago-based Gas Research Institute (GRI), which in 1997 tested 96 CO detectors of different types, and concluded that electrochemical detection technology gave the best performance. The GRI's report expressed concern that the existing North American safety standard (UL2034, set by Underwriters Laboratories (UL), an independent safety testing and certification organisation based in Northbrook, Illinois, US) was not strict enough to ensure acceptable performance of detectors.

Partly due to this concern, the American

Gas Association has created the International Approval Services (IAS) 696-98 standard, which is designed to reduce false alarms and false negatives, and to increase reliability, accuracy and repeatability. So far, says Kerr, the only CO detector to meet the IAS 696-98 standard is that manufactured by his company. This detector has a patented self-testing mechanism that produces a small burst of hydrogen, by electrolysing moisture from the air, to test the sensor every 24 hours. The detector responds to hydrogen in a similar way to CO, so this system actually checks the operation of the sensor, unlike conventional test buttons that only test the alarm circuits. If the sensor fails to respond to the test, a malfunction alarm is sounded.

However, even these detectors are not the end of the story. George Kerr told *Chemistry in Britain* that Aim Safe-Air Products is currently collaborating with a British company, Analox Sensor Technology, to develop new CO sensors based on more advanced technology.

Other applications

The domestic detectors described here represent some of the most visible uses of chemical sensor technology – to the general public, at least. But other gas detectors are available that can measure levels of many different gases and vapours, including hydrogen sulphide, nitrogen oxides, sulphur dioxide, ozone, phosphine and hydrazine. Such detectors have many industrial and commercial applications, such as measuring levels of pollutants in exhaust gases, monitoring hazardous waste sites, and checking air quality in factories and offices. On spacecraft too, any build-up of toxic gases is potentially disastrous – as anyone familiar with the fate of *Apollo 13* will know – so an early warning system is especially important.

For most of us though, smoke and carbon monoxide are more likely to be a worry. Fitting good quality alarms – and ensuring that they are properly installed and maintained – could be a lifesaver. □

ℹ Information

- G. Troughton, *Platinum Metals Review*, 1998, 42 (4), 144
- R. W. Bukowski and R. G. Bright, *Fire Journal*, September 1976
- *How stuff works*: www.howstuffworks.com/smoke.htm
- RoSPA's website: www.rospa.co.uk
- AIM Safe-Air Products' website: www.aimsafeair.com
- Quantum Group's website: www.qginc.com
- Hamel Volunteer Fire Department's 'CO frequently asked questions': www.freenet.msp.mn.us/people/guestb/pubed/cofaq.html
- Robert W. Cattrall, *Chemical sensors*, Oxford: OUP, 1997.
- The Gas Analysis and Sensing Group (GASG) has produced a directory of gas sensor research at UK universities. The 97 page directory is available, price £95 (£45 for universities), from: Dr J. Watson, Chairman, Gas Analysis and Sensing Group, Department of Electrical and Electronic Engineering, University of Wales, Swansea SA2 8PP, UK; tel: 01792-295414; fax: 01792-295686; e-mail: j.watson@swansea.ac.uk

RS•C

The nitrogen oxide story

Teachers' notes

Objectives
■ For students to become familiar with handling information on the student safety sheets.

Outline
This activity has been included as an exemplar of how student safety sheets can be used and it is hoped that the approach will be adopted and used with other chemicals.

Teaching topics

This activity is suitable for 11–14 year olds. It can be used as a stand-alone piece of work. Some teachers may wish to use this to introduce the idea of concentration, stressing that the more concentrated samples of nitrogen oxide will do more damage than less concentrated samples. It could also be used when teaching about elements and compounds.

Teaching tips

This activity could be introduced by discussing some of the material on the student worksheet, such as the meaning of the hazard symbols and the difference between the oxides of nitrogen. This could lead into further discussion about the contexts mentioned on the student sheets. The following points should be noted:

■ Many students suffer from asthma and will know that things that irritate the respiratory system have the potential to trigger an asthma attack.

■ Not all students will have heard about Alzheimer's disease or Parkinson's disease.

■ Not all students will know that nitroglycerine is an explosive.

After a discussion, the student sheets could be set as a homework exercise.

Timing

60 minutes

Answers

1. Toxic – skull and cross bones symbol.
2. Initially it would irritate mouth, throat, and respiratory system. If too much (more than 44 mg m^{-3}) was breathed in the person would be poisoned.
3. (b) a girl who has asthma.
4. Evacuate the laboratory, and open the windows to allow good ventilation.
5. N_2O (laughing gas) is safe to breathe in. It is used as an anaesthetic.
6. $N_2(g) + O_2(g) \rightarrow 2NO(g)$ or word equation.
7. NO is a contributor to acid rain and photochemical smog.
8. Use of a catalytic converter, so the engine can run at lower temperatures, reducing the formation of NO.
9. Not very good, a toxin and polluter.
10. Nitroglycerine is an explosive. It has been used to make bombs.
11. During the treatment of heart disease, nitroglycerine releases small amounts of NO into the body. At last NO is being used for something good.

The nitrogen oxide story

Carefully study the safety information for nitrogen oxides to find out what sort of reputation nitrogen oxide really has.

Nitrogen oxides

Substance	Hazard	Comment
Nitrogen (mon)oxide (NO) gas	VERY TOXIC	Very toxic if breathed in. Irritates eyes and respiratory system. For 15-minute exposure, concentration should not exceed 44 mg m^{-3}. Reacts with oxygen in atmosphere to form nitrogen dioxide (see below). May form by the reaction between oxygen and nitrogen in air, especially in car engines. This is a major contributor to acid rain and photochemical smog. The mixture of NO and NO_2 formed in this way is often referred to as NO_x.
Nitrogen dioxide (NO_2) and dinitrogen tetroxide (N_2O_4) gas	VERY TOXIC CORROSIVE	Very toxic if breathed in. May cause dizziness, headaches and coldness. Irritates eyes and respiratory system. Serious effects may be delayed until after apparent recovery. May trigger asthma attack. For 15-minute exposure, concentration should not exceed 10 mg m^{-3}. Formed as air pollutant from nitrogen monoxide (see above). Formed in laboratory by action of heat on many nitrates, and by nitric acid on some metals. Very soluble in water – risk of suck-back.
Dinitrogen oxide (N_2O) "Laughing gas"	LOW HAZARD	Anaesthetic in large amounts. Used as general anaesthetic eg by dentists.

Questions

1. Which safety symbol would you use to describe nitrogen oxide (NO) gas?

2. What would happen to you, if you breathed in NO?

3. Which person would be affected the most by a leak of NO gas into the air: (a) a boy with a broken leg, (b) a girl who has asthma, or (c) a girl who has measles?

4. What action should be taken if there was a NO leak in the laboratory?

5. Is it safe to breathe in any of the oxides of nitrogen? Give a reason for your answer.

6. Write an equation for the formation of nitrogen oxide in car engines.

7. Why do you think NO is considered to be a major pollutant?

8. What could be done to reduce the amount of NO given out by vehicle exhausts?

9. What sort of reputation do you think the small gaseous NO molecule has?

In 1977, things started to look up for NO… an announcement was made by Ferid Murad of the University of West Virginia…

Nitroglycerin (another molecule, that used to suffer from bad image problems), commonly used to treat heart disease, releases nitrogen oxide in the body.

10. Find out why nitroglycerine suffered from bad image problems.

11. Why was Ferid Murad's announcement good news for NO?

What happened during the next 20 years?

After the announcement was made by Ferid Murad, scientists were a bit confused and excited at the same time. They had many questions to ask such as:

- What does NO do in the body?

- Could it be the answer to heart problems?

- How much is released?

- Why is it not killing the cells?

- Could NO be the answer to some medical conditions.

A big investigation was carried out and here are some of the results:

- NO instructs blood vessels to relax as they respond to nerve messengers.

- White blood cells use NO to kill bacteria and other disease causing organisms.

- Sometimes the white blood cells get over active and produce too much NO which causes the blood vessels to dilate too much and the patient suffers shock.

- NO is active in the brain and could be responsible for memory. This could help scientists find a cure for Alzheimer's and Parkinson's diseases.

- Viagra, the drug that cures some impotent men, owes its performance to NO.

In 1998, the Nobel Prize for Medicine was awarded to three men, Louis Ignarro, Ferid Murad and Robert Furchgott for the research done into the role of NO in living organisms.

During the 1980s and 1990s scientists discovered the good side of the NO molecule. NO is really a medical wonder molecule. However, most people still regard it as a health hazard. Well, in fact they are right. When you have enough NO around, it is still a pollutant.

12. The NO molecule wants to improve its image. Your job is to design a flyer to market NO and help NO lose its bad reputation.

RS•C

Radiation doses

Teachers' notes

Objectives
- To understand the risks associated with radiation.

- To realise how the risks associated with radiation compare with other risks associated with other areas of life.

- To learn how radon levels are controlled.

Teaching topics

This activity is appropriate for 12–14 year olds. This piece of work can be used when teaching about the Periodic Table, as its main focus is around the risks associated with radiation and in particular, radon. The health risk posed from radiation has been put into context with risks associated with other jobs.

Before reading these notes look at the student worksheets.

Background information

Radon reduction methods
High radon levels in homes and other buildings can usually be reduced by simple technical means.

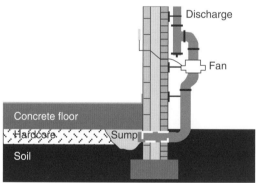

A small underfloor sump is an effective way of preventing radon from entering a building. Air and other gases in the soil are extracted by a low power fan through a pipe under a solid floor. The sump can be inside or outside the building, whichever is most appropriate.

Sumps generally cause little change to the external appearance of the building. This is the extract pipe from a radon sump. It is about 10 cm in diameter.

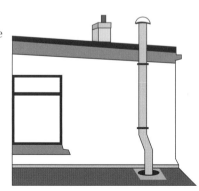

RS•C

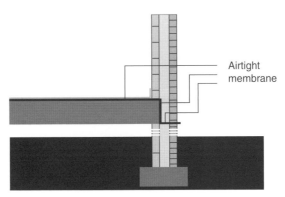

Airtight membrane

For new houses, simple measures can be taken cheaply during construction to prevent high radon levels. This diagram shows an airtight membrane across the floor and through the walls.

Blowing air from the loft into the house can be effective when only a small reduction in radon level is required.

(Reproduced with kind permission of NRPB from the Radon At-a glance series, 1994.)

Sources of information

Information on all aspects of radiation protection is available from the National Radiological Protection Board (NRPB) and is published as a series of 'At-a-glance' leaflets, which make good, informative wall charts.

Living with radiation (5th Edition), Didcot: National Radiological Protection Board, 1998.

The 'At-a-glance' leaflets are free and can be obtained from:

NRPB, Chilton, Didcot, Oxon OX11 ORQ
Tel: 01235 831600
Email: nrpb@nrpb.org.uk
Web: **http://www.nrpb.org.uk** (accessed June 2001)

RS•C

Answers

Radiation doses

1. Radon from the ground – natural; gamma rays – natural; medical – artificial; food and drink – natural; cosmic rays from the atmosphere – natural, other – artificial.

2.

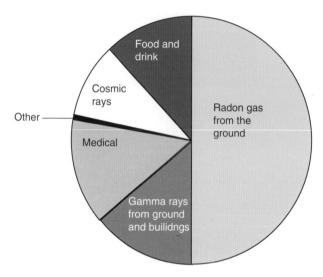

3. Radon

4. Radiation that produces ionisation in matter. When these radiations pass through the tissues of the body, they have sufficient energy to damage DNA.

5. Coal mining

6. Being hit by falling objects, slipping and falling off scaffolding etc. Accept any reasonable answer.

7. Service industry

8. 1 in 17000 – safer than some other industries.

9. 1 in 7700

10. No, because the average risk of death in the UK from nuclear discharge is 1 in 140000, which is very low compared to heart disease, smoking, and accidents in the home and on the road.

11. Yes. If people were not protected from radiation, there would be a greater of risk of dying from radiation. It is essential to protect people who work with radiation everyday and those who live in a high radon area.

What is radon?

12. Rn

13. 8

14. Gas

15. No – group 8 of the periodic table contains the noble gases.

Where do we find radon?

16. The Southwest peninsular, The Derbyshire uplands, Northamptonshire, The Mountains of Mourne (Northern Ireland), The Dee Valley, Northeast Scotland.

17. The Southwest peninsular (fractured granites), The Derbyshire uplands (limestone), Northamptonshire (sandstones and limestones), The Mountains of Mourne (granites), The Dee Valley (granites and limestone), Northeast Scotland (limestone).

18. Radon can get into buildings in a number of ways, including through concrete floors, construction joints, cracks in walls, cavity gaps, stone walls and timber floors. Inside the buildings, the air pressure tends to be lower than outside, so radon is drawn in through the gaps. This is due to the wind blowing causing the Venturi effect.

19. Highest in January, lowest in July or August.

20. Pupils should appreciate that the indoor radon levels will be lower when the doors and windows are open. The radon concentration will therefore be lower during the summer months. The radon levels build up overnight when the building is shut up.

21.

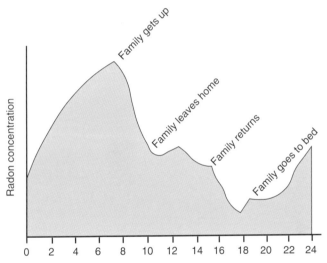

22. Short-term measurements of radon concentrations in air may be misleading because of the way buildings are heated and ventilated. Factors affecting the daily levels will depend upon the time of the year, the weather, the activities of the people living or working in the building, for example if the building is empty at the weekend.

The radon in the ground is drawn in by the air movement found in most buildings. This is caused by warm air inside the building rising and escaping through the roof (this is why hot air balloons work) and the effect of wind blowing across eaves, over chimneys and around houses (variations of the venturi effect). Even assuming a constant indoor temperature and ventilation regime, the variability of the temperature outdoors and changes in the wind speed and direction will cause variations in the indoor radon concentration.

23. There are so many variables involved in the measuring radon levels that it would be difficult to exactly reproduce graph A. The general trends would be similar.

24. See the background information for teachers (p. 46–47).

25. The bar chart should show lifetime risk and the radon level in either unit.

26. Advise them about the risk of lung cancer and suggest that they give up smoking or at least try to reduce the number of cigarettes per day. Also, to get the NRPB to measure the radon level in their house and depending on the result take action to lower the radon level.

RS•C

Lifetime risk %	Radon level Bq/m^3
0.1	20
0.5	100
1.0	200
2.0	400

← Action Level

Table 3 Lifetime risk of lung cancer potentially induced by non-smokers
(Source: NRPB, 2000.)

If you smoke 15 cigarettes a day, you can multiply the risk by a factor of 10, so for example at the Action Level your risk is 10%.

RS•C

Radiation doses

Many people are scared of nuclear waste, because it gives off radiation and too much radiation can cause some types of cancer. But did you know that we are all exposed to radiation, everyday? The radiation dose depends on where we live, where we work and even on what we eat and drink! Nearly all of the radiation comes from natural sources. Radioactivity is commonly measured in bequerels – decays per second, but radiation dose is measured in sieverts, which takes account of how the different types of radiation cause harm to our bodies.

The average yearly dose to the UK population is 2.6 mSv (millisieverts). The information in the table shows where this dose comes from:

Source	Amount	Natural / Artificial
Radon from the ground	50%	
Gamma rays from the ground and buildings	13.5%	
Medical	14%	
Food and drink	10%	
Cosmic rays from the atmosphere	12%	
*Other	0.5%	

*Includes: Nuclear discharge <0.1%, Products <0.1%, Fallout 0.2%, Occupational 0.3%.

Radiation sources
(Source: The National Radiological Protection Board, 2000.)

1. Decide whether each source listed in Table 1 is natural or artificial.

2. Using the information in Table 1, draw a pie chart to showing the sources of radiation.

3. Which is the biggest source of radiation?

There are two types of radiation; ionising radiations such as cosmic rays and x-rays and non-ionising radiation such as light and heat, radio waves and microwaves.

4. What do you think is meant by ionising radiation?

Both types of radiation can be very useful but some can cause cancer if you are exposed to too much of it. So just how risky is radiation?

RS•C

Average annual risk of death in the UK from industrial accidents and from cancers due to working with radiation		Average risk of death in the UK from some common causes	
Coal mining	1 in 7000	Smoking 10 cigarettes a day	1 in 200
Oil & gas extraction	1 in 8000	Heart disease	1 in 300
Construction	1 in 16000	All cancers	1 in 400
Radiation work (1.5 mSv y^{-1})	1 in 17000	All causes for a 40 year old	1 in 700
Metal manufacture	1 in 34000	All radiation (2.6 mSv y^{-1})	1 in 7700
All manufacture	1 in 90000	Accidents in the home	1 in 15000
Chemical production	1 in 100000	Accidents on the road	1 in 17000
All services	1 in 220000	Radiation dose (1 mSv y^{-1})	1 in 20000
Nuclear discharge (0.14 mSv y^{-1})	1 in 140000	Pregnancy, for mothers	1 in 170000

Risk of dying in the United Kingdom

(Source: *Living with radiation* (5th Edition), Didcot: National Radiological Protection Board, 1998.)
Note: Radiation workers also run the risk of death from more conventional causes at work.

5. Which jobs put workers at the most risk?

6. What are the risks associated with working in the construction industry?

7. Which jobs offer the least risk to their workers?

8. What is the risk of a radiation worker dying from cancer?

9. What is the chance of the average person dying from radiation exposure in the UK?

10. Do you think that nuclear discharge poses a real threat to people living in the UK?

11. Do you think that it is necessary to protect people from radiation?

Remember that most radiation comes from natural sources, and so it is difficult to control it. It does, however, seem unfair if some parts of the country have more radiation than others. We are now going to investigate the biggest source of natural radiation, radon.

What is radon?

Use the Periodic Table to answer the following questions:

12. Radon is an element. What is the chemical symbol for radon?

13. Which group of the Periodic Table is radon in?

14. At room temperature is radon a solid, liquid or gas?

15. From your knowledge of the Periodic Table, would you expect radon to be reactive? Give a reason for your answer.

Where do we find radon?

Key

Affected areas where 1% or more of homes have radon levels above the Action Level of 200 becquerels per cubic metre

Edinburgh

Belfast

Dublin

Manchester

Northwich

London

Exeter

A map showing how the amount of radon gas varies throughout the United Kingdom
(Reproduced with kind permission of NRPB.)

The National Radiation Protection Board (NRPB), advises that people living in areas of high radon concentrations should take precautions to protect themselves and their houses.

16. Which areas in the country suffer from high radon levels?

17. Radon comes from uranium that occurs naturally in all rocks and soils. When the rocks are particularly porous, such as limestones and sandstones, and / or contain more than the average amounts of uranium, such as some granites, the risk of high levels of radon is greater. Use the map above to predict where large amounts of granite or limestone may be found in the UK.

How big is the radon risk?

Outside, radon quickly disperses in the atmosphere and causes no risk, but indoors the radon levels build up and so does the risk of lung cancer. The yearly average radiation dose from radon is 1.3 mSv, with a range of 0.3–100 mSv.

18. In what ways could radon get into the house?

Graph A

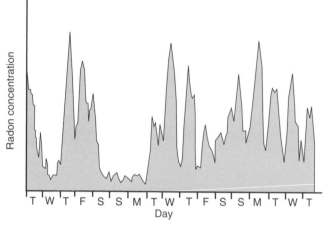

Graph B

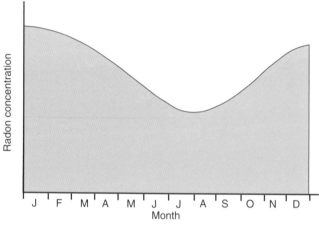

Graph C

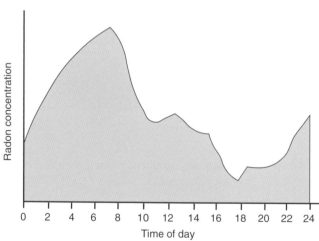

Graphs to show how the concentration of radon indoors varies
(Reproduced with kind permission of NRPB.)

19. Use graph B to determine in which months the radon levels are highest and lowest?

20. Suggest a reason for your answer to question 18.

21. Graph C shows how radon levels vary from hour to hour. At what time does the family:
    a) get up
    b) go to bed
    c) go out
    d) get back home?

22. Look at graph A, and suggest why the weekly radon levels do not show a regular pattern each day.

23. How reliable do you think the data in graph A are?

24. Suggest how the amount of radiation indoors could be reduced.

Health studies from around the world have linked radon and lung cancer. The lung cancer caused by radon proceeds in exactly the same way as the cancer caused by smoking.

NRPB has estimated the risk of lung cancer from lifetime exposure to radon in the home. The average annual radiation dose from radon is estimated to be 1.3 mSv, and this poses a 0.3% or 1 in 333 risk of getting lung cancer. However, if you smoke fifteen cigarettes a day, you can multiply the risk of getting lung cancer by a factor of 10. In 1998, a joint study between the NRPB and the Imperial Cancer Research Fund concluded that about 5% of lung cancers in the UK are caused by radon in the home.

The unit used to measure radon levels is the becquerel per cubic metre ($Bq\ m^{-3}$).

1 becquerel (1 Bq) = 1 atomic disintegration per second.

Lifetime risk %	Radon level, $Bq m^{-3}$	Annual radon level, mSv
0.3	20	1.0
1.0	70	3.5
2.0	130	7.0
3.0	200	10.0

The lifetime risk of the general population getting cancer
(Source: NRPB, 2000.)

25. Draw a bar chart to show how the risk of getting lung cancer changes as the amount of radon increases.

26. What advice would you give a smoker living in a high radon area?

Extension work

Find out the advice the NRPB gives to those people living in high radon areas. A good place to start your research is the NRPB website at **http:// www.nrpb.org.uk** (accessed June 2001).

RS•C

References

1. D. Warren, *Green Chemistry*, London: Royal Society of Chemistry, 2001.

2. M. Burgess, *Living Dangerously*, published to accompany the Channel 4 Equinox programme, (ISBN 1 85144 2529).

3. *Risk, analysis, perception, management*, London: The Royal Society, 1992.

4. *Green Chlorine*, York: Chemical Industry Education Centre, 1996.

RS•C